Heat Treatment Secrets for Bladesmithing

Table of Contents

Introduction

Steel has more than 95% iron with trace amounts of Carbon.

Why is this important? It is because pure iron is a comparatively soft element that can easily rust. On the other hand, with the right alloy, it turns into a material that is difficult to bend.

Steel can be made to be strong, more solid than iron, and resistant to corrosives. It is also a versatile material, which you can use to make objects ranging from your everyday kitchen utensils and tools to office appliances to automobile parts, building materials, and machinery.

But you might be curious; aren't there any materials better than steel? There are. For example, carbon fiber, made from thin sheets of carbon, is considered to be more durable, flexible, heat and electricity resistant, and stronger than steel. However, it cannot be used more than steel. That is because of the availability of materials.

You see, among all the metals that we currently know of, iron is the second most naturally abundant material found on Earth after aluminum. About 4.7% of the Earth's crust is made up of iron and it appears as different types of oxides. The main product that iron is used to manufacture is, you guessed it, steel.

Apart from the easy access to iron, one of the things that makes steel important in today's manufacturing and construction landscape is the fact that steel can be recycled easily. In fact, the American Iron and Steel Institute has discovered that nearly 88% of steel is recycled in many parts of the world. One of the more spectacular aspects of recycling steel is its ability to retain some of its properties. If you start recycling other materials, they begin to lose their strength after each recycle. But not steel. Even after recycling multiple times, steel does not lose its integrity and strength. This feature also becomes important for the simple fact that you do not see a lot of wastage when it comes to steel.

So what does this have to do with bladesmithing?

Simple, you are working on one of the most magnificent materials known to man. When you apply heat treatment to this material, you are not destroying its strength. You are not causing it to rust. You are not deteriorating its quality.

You are making use of its advantages to make some incredible tools for yourself.

Heat treatment is a simple but effective technique that prevents your blade from shattering.

However, before we get down to working with steel, it is important to know just what exactly we are doing.

Chapter 1: Iron Management

It's Gettin' Hot in Here! What is Heat Treatment?

Essentially, no material or finished product can be manufactured without sending it through the process of heat treating. In this process, a particular metal is heated to a high temperature and then cooled under specific conditions to improve its characteristics, stability, and performance.

Through heat treatment, you can soften a metal, which allows the metal to become more flexible. You can also use heat treatment to harden metals, ensuring that their strength is improved.

In fact, heat treatment is essential if you are in the business of manufacturing parts for automobiles, aircraft, computers, and other forms of heavy machinery and tools. In other words, if you want something important built, then you need to subject the material to heat treatment.

Iron, and more specifically steel, are the most common materials that go through heat treatment. However, that does not mean that other materials cannot be treated with heat.

It has been estimated that heat treatment adds more than $10 billion in value to materials by enhancing their durability and quality.

In short, this process is quite important and you are going to learn to use it.

Temper Tantrums

One of the processes of heat treatment is called tempering. In this process, you are basically altering the mechanical attributes (usually the flexibility and strength) of steel or products and items made from steel. Tempering releases the carbon molecules confined in the steel to diffuse from martensite. Martensite is a form of a crystalline structure consisting of brittle carbon that exists in hardened steel. Because of martensite's features, the steel may be hard, but it also becomes brittle, rendering it useless in most applications.

It then allows the internal stresses that may have been formed due to past uses to be discharged from the steel. This results in the alloy becoming more durable.

So how does one temper their steel? Firstly, the steel is heated to a high temperature, but it is not allowed to heat up beyond its melting point. Once that is done, it is then cooled in air. There is no fixed temperature for all forms of steel. They each have their own temperature range that must be reached first. Alternatively, temperatures can also be adjusted to fit the degree of hardness one would like to reduce in the steel.

For example, when you heat the steel at low temperatures, then you are reducing the brittleness in the material and increasing the hardness. This could be applicable for certain purposes, such as setting the steel.

Other times, the steel is heated at high temperatures. When this is done, it reduces the toughness of the metal but then increases its flexibility. This is useful when you would like to work further on the metal.

When you temper the metal, it is important to heat the metal gradually until it reaches the temperature you would like to work with. This prevents the metal from cracking.

Annealing

Annealing is another process similar to heat treatment, focused on softening the metal or reducing the hardness of the material. We usually use this process after subjecting the material to a softening or hardening treatment. Through annealing, you are trying to return materials to a state before they were cooled. This allows you to reshape them or form them in a particular manner with relative ease.

When metals are subjected to cold treatment, they could become so hard that if you perform any more work on them, they might begin to crack. Obviously, that is not something we would like to achieve. When you perform the process of annealing before you subject the metal to cold treatment, then you reduce the chances of cracking the metal.

Typically, you might discover that annealing takes place in large ovens. The space inside the over must be large enough that when the metallic object is placed inside, there is still room for proper air circulation around the metal.

When you cool materials, their degree of malleability reduces. Their strength goes up and it becomes difficult to work on them again.

When you use annealing, you are giving yourself the opportunity to perform modifications on the material, should you require it.

In this process, the metal is heated to a temperate where it is possible to attain recrystallization. This means that new non-deformed grains take over the positions of the deformed grains. And what exactly are grains? In metallurgy, each grain is a single crystal that consists of a specific arrangement of atoms. When you have deformed grains, then you cannot work on the metal without causing more deformity. In this case, the deformity appears in the form of cracks. When you perform the annealing process, you are forming new grains, which means you are giving yourself the ability to work on the metal again.

Quenching

This is a pretty popular process in metallurgy and you might have seen it being performed on TV or in the movies. Quenching is an essential step for heat treatment. Without it, the other steps are rather futile. Basically, the metal worker takes the steel and then dips it into a cauldron, often creating a dramatic effect. You can see steam rising and the water sizzling as the metal cools. However, water is not the only medium that the metal worker uses for the process of quenching. Before we dive into that, let us look at the quenching process in more detail.

Quenching is a process that occurs after another process where the metal is heated to high temperatures. Examples of the kind of processes that precede quenching are annealing (which we looked at earlier) or normalizing (which we shall look at in the next section). In both annealing and normalizing, the cooling process can take some time. This could affect the strength of the steel, causing it to lower more than necessary. Through quenching, you are lowering the temperature considerably, which could benefit the work that you are doing. Metal workers usually apply this method so that they can prevent the cooling process from altering the molecular structure of the metal.

Quenching is typically done by subjecting the metal immediately to a certain liquid, typically water, or to forced air. In a forced air cooling system, the air is pushed out through specially arranged ducts that help in, you guessed it, cooling the temperature of an object quickly. The water or air used for the process of quenching is termed the "medium."

But you have to be careful not to plunge the hot steel into the medium too quickly, it can ruin the quality of your heat treat. This is a huge mistake that you have to avoid.

Another warning, never use a plastic container to store your quenching medium. Doing so is a sure-fire method to start a fire in your workshop.

Now you might think to yourself - is there any other liquid that can be used for quenching? That is a valid point actually. We are so used to seeing the movies show metallic objects such as swords and weapons being dipped into water that we are not aware of any other liquids that can be used. However, here a couple of other liquids that are used for quenching:

Oil

There are numerous oil options that you can use for the process of quenching. You have fish oils, vegetable oils, and certain mineral oils that can help you attain the desired effect. With each medium, you have a different rate of cooling. When you use oil, you are using a liquid that has a higher cooling rate than air but cools the metal down more slowly than water.

When choosing oils, here are a few options that you can consider for your bladesmithing requirements:

Food Grade Oils

Many bladesmiths and knife makers utilize food grade oils for the sole reason that they are cheaper and readily available. You can head over to your local supermarket or store and find one on their shelves. In fact, you are also spoilt for choice, with the number of different brands available to you. A famous food oil used in quenching is canola oil.

Canola oil is very easy to find. Just a visit to your grocery store will be enough. It also reduces the chances of your steel cracking while you quench it, as it doesn't allow your steel to cool too quickly.

During the quenching and tempering processes, food oils spread a much more bearable odor than other types of oil. While you might think that this is a minor point to make note of, it might affect you if you have a workshop attached to your home or within your home.

To quench in food grade oils, preheat the oil to anywhere between 150°F to 200°F.

Motor Oil

Another oil that is popularly used in the knife-making industry or hobby is motor oil. People use both new and used motor oil, depending on their requirements. The advantage of motor oil is that it is really cheap to obtain. In fact, used motor oil is free; you may have it in your garage, you might find some in your friend's garage or even in the local store.

However, do note that used motor oil tends to leave a stench. Additionally, you might find a dark film that coats the blade you are working on which is quite difficult to remove. Another reason why you might want to stay away from used motor oil is the fact that they contain quite a few toxins that could be potentially harmful when inhaled. For beginners, it is highly recommended that they do not use motor oils. Experienced knife makers who have worked with the blade for many years might be able to spot a good motor oil. Nevertheless, they prefer to avoid it as well if they have a choice.

To quench in motor oils, preheat the oil to anywhere between 200°F to 250°F.

Mineral Oil

Mineral oil is an alternative to motor oil. Their benefit is that they do not give off any odor, do not contain any harmful contaminants or toxins, and some of them are fairly odorless. If you can get your hands on high-grade mineral oil, then you might be able to avoid the flames that flare up during the quenching process.

In many cases, mineral oil is recommended for beginners.

To quench in mineral oil, preheat the oil to anywhere between 250°F to 300°F.

Baby Oil

You heard that right. Many knife makers head over to their local store and get a lot of baby oil for the process of quenching. Baby oil contains minimal contaminants, does not create flames easily, and tends to give rise to as little odor as possible. Should you prefer, you could even get yourself a scented baby oil. You can quench your blade and leave behind a nice smell as a bonus!

The downside is that, as you might require a fair amount of baby oil, you might find yourself shelling out a bit of cash to acquire the right volume of oil for your project.

You can use the same preheating temperature for baby oil as you did for food oil, between 150°F to 200°F.

Quenching Oils

Finally, you can purchase special quenching oils for your project, taking advantage of the fact that you can find the right oil for a specific purpose. Working on a blade? You have a quenching oil for that. Working on an anvil? Sure, there is a quenching oil for that. Need a specific quenching oil for a specific steel? Sure, no problem.

Of course, with the number of options available to you, quenching oils might be a bit more expensive than other forms of oils.

For quenching oils, the quenching temperature is unique to the type of oil you choose to purchase. The quenching temperature will be mentioned on the packaging, allowing you to not only have instructions for quenching but also to help you decide just what quenching oil you would like to use for your project.

Brine
When you dissolve rock salts in water, then the result is a liquid called brine. What is so special about this brine? It lowers the concentration of atmospheric gases. This, in turn, lowers the number of bubbles formed during the quenching process. The result is that brine cools the metal faster than water can.

Here is a key point to remember about

quenching. You need to use the medium that best suits your purpose. Different kinds of steel might require different periods of cooling, even though your main aim is to cool them as fast as possible.

When you want to quench in brine, you typically keep the temperature between 150°F to 200°F. However, in some cases you can bring the temperature to around 100°F as well, depending on the steel.

Having understood the different kinds of oils, a few people often ask, "Is there a method of quenching that is ideal for the metal?" As a matter of fact, there is. When you are quenching, make sure that you move the blade forward and backward in the liquid, almost as though you are poking something. This will allow the metal to cool faster.

Normalizing

During normalizing, you are refining the size of the grain in the metal. Should you need a refresher course on what grains are, you can always refer to the section on annealing. After normalizing, the mechanical properties of the metal are improved.

Normalizing sets a uniformity to the structure

of grains in the metal. After you have achieved uniformity, you have reduced the degree of deformity of the metal. This allows you to get an even finish and a wonderful product in the end.

In the process of normalizing, steel is heated to a really high temperature and then cooled by leaving the metal at room temperature. This process of rapidly heating up the metal and then slowly cooling it down makes changes in the microstructure of the metal, making it elastic and durable. Normalizing is almost like a process of correction. This is because it is typically used when some other process unintentionally increases the hardness but decreases the malleability of the metal. What makes normalizing different from other processes such as annealing is that it uses room temperature to cool down the metal, rather than any medium or special technique.

We all know the basic property of heat; it expands. This is practically what happens to metals as well. The heat causes them to expand. It also affects the structure, magnetism, and electrical resistance of the metal. What you are doing is making sure that the metal turns out just the way you want it to. The temperature to which you heat the metal and then the temperature to which you cool the metal affects the overall properties of the metal. As seen above, we can affect the metal's hardness, malleability, ductility, and even its resistance to corrosion by using the right heat treatment.

Chapter 2: Real Steel

In this chapter, we are looking at the properties and temperatures of the various types of steel you will be working with. We are going to deal with as many details as possible for each of the steel variations. So let us start with the first one.

1080 steel

1080 is part of a series of steels that are known as carbon steels. They are called so because these variations of steel possess carbon as the principal alloying element. Now do not confuse alloying element with the primary element (also known as the base element). As we pointed out before, the base element in steel is iron. That does not change. Alloying elements refer to those materials that combine with the base material to form an alloy.

In addition to carbon, this series of steel also has trace amounts of silicon (0.4%) and manganese (1.2%). You might also find small quantities of other elements such as molybdenum, chromium, copper, nickel, and aluminum. This form of steel has one of the highest electrical conductivities among molded carbon or non-alloy forms of metal. In addition, it has comparatively low ductility. In some cases, you might also find this metal to have a moderately high tensile strength. The below information will allow you to understand more about 1080 carbon steel in detail.

Chemical Composition
The main elements present in 1080 steel are highlighted below;

Iron - around 98%

Carbon - around 0.80%

Manganese - around 0.60% to 0.90%

Sulfur - a maximum of 0.05%

Phosphorous - a maximum of 0.04%

Mechanical Properties
Elastic modulus - 190-210 GPa

Let us try to understand what elastic modulus really means. Have you ever stretched a spring? You might have tried to pull it to its maximum length and then let go. Typically, the spring bounces back to its original position. However, if you were to keep stretching the spring, then you would reach a point where the spring will not return to its original form. This property of elasticity holds for many materials, including steel.

When you apply a certain amount of force to a material, then you stretch the material. When you remove the application of force, then the material returns to its original shape. However, there is always a limit to the amount of force you can apply to stretch a material. The ratio of strain that an object can handle because of its elasticity is called the elastic modulus of the material.

Poisson's ratio - 0.27 to 0.30

When a material stretches, it does so in two different directions. If we were to take the example of the spring, then upon stretching it, it increases in length but decreases in diameter. The increase in the length is known as the longitudinal strain and any changes to the diameter or width of the material are referred to as the lateral strain.

It is this change in the lateral strain and longitudinal strain that you denote using Poisson's ratio.

Additionally, carbon steel is also available in two different formats.

You have carbon steel that is hot rolled and you have carbon steel that is cold drawn. Let us look at the difference between the two.

1095 Steel

Once again, 1095 steel forms a part of the carbon steel series. This means that carbon is the major allowing steel here once again. Here are the properties of 1095 steel.

Chemical Composition
Iron - around 98%

Carbon - around 1.03%

Sulfur - less than or exactly 0.050%

Phosphorous - less than or exactly 0.040%

Manganese - less than or exactly 0.50%

Physical Properties of 1085 steel
Density - 7.85 g/cm³

Melting point - 2759°F (1515°C)

Mechanical Properties
Tensile strength - 685 MPa

When you use the term tensile strength, then you are referring to the degree to which a material can stretch before it will break or fail. Now, tensile is different from elastic modulus because it does not measure the rate at which a material can bounce back to its original position after stretching.

The final tensile strength of a metal is measured by taking the area of the material, which is also the cross-section of the metal. You then divide it by the stress exerted on the metal. The stress that metal bears is typically represented as pounds or tons per square inch of metal. The tensile strength of a metal is an important measurement. This is because it looks at a metal's ability to function in various applications. It is a widely used measurement for metals and alloys, so it would be prudent to make a note of it when working with various metals.

Yield strength - 525 MPa

Elastic modulus - around 210 GPa

Poisson's ratio - around 0.30

15N20 Steel

As you are making blades, 15N20 is an incredible steel for the purpose. What makes it a standout steel is the fact that it is tough and allows you to subject it to heat treatment easily. One of its more common uses is to make the bandsaw blades that are commonly found in sawmills.

One of the remarkable things about 15N20 is that it can be used as a blade material on its own. However, should you choose to, you can mix with other steel materials in pattern welded damascus.

Now you might be thinking. What exactly is damascus? Well, when you make steel products with a wavy pattern on the surface, then you have a damascus steel. You can achieve this form of steel by using a hammer to weld strips of iron and steel. Then you have to subject both metals to repeated forging and heat treatment processes. Damascus steel is used mainly for the purpose of making blades for knives and swords. We will look at a popular combination of damascus steel further down below.

As nickel is also present in 15N20, it provides a nice contrast to the steel, giving it a wonderful aesthetic appearance. Once you have made the blade with 15N20 steel, it also becomes easy to sharpen. You can create a really fine edge with the steel and also make it tough.

Apart from the above, nickel has many uses when it is present in the steel.

Chemical Composition

Iron - approximately 97%

Carbon - approximately 0.75%

Manganese - a maximum of 0.4%

Nickel - approximately 2.0%

O1 Steel

There are certain types of steels known as cold-work steels. These are high carbon steels and typically appear in three categories. These categories are:

- Oil-hardening steels

- High-carbon and high-chromium steels

- Air-hardening steels.

Among the three of the aforementioned steels, we are going to focus on oil-hardening steels. These are also known as O steels and come in various forms. The one that we shall be working with will be O1. So let us look at what this steel is made of.

O1 steel's main composition consists of metals such as chromium, manganese, and tungsten, and is relatively inexpensive. The following datasheet provides details about O1 type steel. Now, we already know that iron is the major component in steel, so let us try and understand what other elements we are going to be dealing with in O1 steel.

Chemical Composition

Carbon - around 1%

Manganese - around 1.4%

Silicon - around 0.40%

Chromium - around 0.60%

Nickel - 0.30%

Tungsten - around 0.60%

Vanadium - 0.30%

Copper - 0.25%

Phosphorous - 0.03%

Sulfur - 0.03%

We can see that O1 has a fair mixture of alloys, with even sulfur and phosphorous coming into the mix. Now it is time to look at some of the mechanical properties and understand a new term present in the measurement.

Mechanical Properties

Density (hardened to 62 HRC) - 7.81 g/cm^3

HRC refers to the Rockwell Hardness of a particular steel. When you are using this term, then the higher the Rockwell hardness number, the stronger the steel. A lower number means that the steel is fairly malleable. With a measurement of 62 HRC, we can see that O1 is fairly strong. If the number were to dip to a level of, say, 32 HRC, then we are looking at a weaker metal which is malleable to work with.

Melting point - 2590°F (around 1421°C)

Damascus Steel

We use "Damascus Steel" to indicate two forms of ferrous materials (those materials that contain iron). These materials can easily be identified by the unique pattern present on them which is produced as a result of a controlled combination of steel and iron. One of the common forms of Damascus Steel includes a combination of 1095 and 15N20 steel. As we saw earlier, 1095 steel is a high-carbon steel, giving you great strength that you can use for your blade. 15N20 steel, on the other hand, allows you to heat treat it really well. When you combine the two, you get a strong and superb blade. Additionally, with 1095 steel's properties, your Damascus blade will be highly resistant to wear over time and after repeated use.

Chemical Properties

Iron - around 97.5%

Carbon - around 0.98%

Manganese - around 0.45%

Sulfur - less than 0.050%

Phosphorous - less than 0.040%

Mechanical Properties

Yield Strength - 485 MPa

Elongation (the degree to which you can extend the length of the blade) - 3.2%.

To understand this, measure 3.2% of the original blade length. That is the maximum additional length you can add to the Damascus steel blade made from 1095 and 15N2o steel when you are extending it.

Stainless steel

Some of the main advantages of stainless steel

might not be obvious. However, when you begin to compare it to other forms of steel, such as plain carbon steel, then the advantages start to become apparent. There are many properties that govern stainless steel. However, the main features that set it apart are:

- High cryogenic toughness

- High work hardening rate

- Low maintenance

- High hardness and strength

- High corrosion resistance

- High ductility

- More attractive appearance

At a molecular level, stainless steel is mainly an iron-based alloy. It also contains a minimum composition of around 10% of chromium. The presence of chromium adds an interesting effect to the entire combination. The chromium adds a protective oxide layer that is also self-healing. It is this layer that confers corrosion resistance to stainless steel. What does the presence of this self-healing feature mean?

Well, stainless steel can essentially retain its

corrosive resistance no matter what fabrication methods you use. This means that despite the work you perform, the healing properties of the chromium layer can always be found in the end product. In fact, even if you damaged the surface while working on it, you will not remove the corrosion resistance abilities of the steel. It will always be there.

Additionally, you might also find that stainless steel has a high cryogenic resistance. This type of resistance is characterized by the fact that the blade or steel retains its toughness at sub-zero temperatures.

Stainless steel also has hot strength. What does this mean? Hot strength is the ability of the steel or blade to retain its toughness at high temperatures. The high levels of chromium allow the steel to have this form of toughness as well.

You may find stainless steels in various forms. Let us look at a few of them.

Austenitic Stainless Steels

These forms of steel are unique because they feature around 6% nickel and more than 16% chromium. They might also contain elements such as molybdenum.

When you include more elements such as molybdenum or even copper in some cases, then you can change the properties of the steel. The type of element used can provide a specific type of property. You can make the steel suitable for dealing with high temperatures or heighten its corrosive resistance. The type of property depends on what you want out of your product.

Additionally, many forms of steel become weak when they are subjected to low temperatures. With the presence of nickel in austenitic stainless allows the metal to retain its strength at low temperatures.

Some of the products that austenitic stainless steel is used in are shown below:

Roofing and gutters

Kitchen sinks

Ovens

Chemical tanks

Doors and windows

Martensitic Stainless Steels

With these kinds of steel, what you are getting is a metal with lower chromium content as compared to other forms of stainless steel. However, you also receive a comparatively higher carbon content.

Some of the products that utilize martensitic stainless steel are noted down below:

Knife blades

Springs

Cutlery

Surgical instruments

Fasteners

Ferritic Stainless Steels

What makes this kind of stainless steel unique is that the major alloying element in them is only chromium. No other elements make a large contribution to the metal as an alloy. The amount of chromium that is present varies between different ferritic stainless steels. Typically, you might find a quantity that ranges anywhere from 10% to 18%. This increases their resistance to cracking from stress corrosion.

Some of the applications in which you might find the use of ferratic stainless steels are:

Fuel lines

Vehicle exhausts

Domestic appliances

Cooking utensils

Duplex Stainless Steels
With duplex stainless steels, you might discover that they include a low percentage of nickel and higher traces of chromium. With the percentages mentioned above, duplex stainless steel features microstructures that give it a close resemblance to both austenitic and ferritic types of steel. The common form of alloy composition in duplex stainless steel includes roughly 23% chromium and around 4% nickel. Alternatively, you can also find close to 22% chromium and around 5% nickel compositions of this steel.

Some of the applications that utilize duplex stainless steel are:

Marine applications

Heat exchangers

Offshore oil & gas installations

Desalination plants

Scrap Steel

Most of us know how to recycle various objects including everyday items like plastic bottles, newspapers, wood and other materials. However, there are many ways in which you can recycle steel as well. The method which is most important to us is in the use of knife making (of course)!

You may find different types of scrap steels. Here are a few common steel scraps that you can get your hands on and their chemical composition:

Leaf Springs

Chemical Composition

Iron - about 98%

Carbon - less than or equal to 0.5%

Silicon - less than or equal to 0.7%

Manganese - around 0.7%

Sulfur - a maximum of 0.05%

Steel Tubes

Chemical Composition

Iron - about 97.5%

Carbon - anywhere from 0.55% to 0.75%

Manganese - around 0.6%

Sulfur - a maximum of 0.035%

Copper - around 0.2%

Chromium - around 0.15%

Molybdenum - around 0.06%

Nickel - less than or equal to 0.20%

Surgical Stainless Steel Items

Chemical Composition

Iron - around 98%

Carbon - around 0.5%

Chromium - around 0.05%

Nickel - less than or equal to 0.2%

Molybdenum - less than or equal to 0.05%

Silicon - less than or equal to 0.04%

Aluminum - no more than 0.025%

Saw Blades

Chemical Composition

Iron - around 98%

Carbon - anywhere from 0.7% to 0.8%

Manganese - around 0.4%

Sulfur - a maximum of 0.05%

Phosphorous - a maximum of 0.04%

Steel Rebar

Chemical Composition

Iron - anywhere from 98.4% to 98.9%

Manganese - anywhere from 0.75% to 1.05%

Sulfur - anywhere from 0.26% to 0.35%

Carbon - around 0.09%

Phosphorous - anywhere from 0.040% to 0.090%

Chapter 3: Working With the Metals

Now we are going to focus on how you can work with each of these materials. Now, what essentially happens is that you must first harden the blade and then temper it properly in order for you to generate the desired product that will be useful for its intended function.

The procedure you have to follow for heat-treating a blade is fairly easy: you have to heat the blade to the right temperature, then you quench it using the correct medium, and then temper the hardness into the metal to a point where is not brittle anymore. That does sound easy enough. Excited to get your metal under heat treatment?

Before you try to harden the steel, identify the type of steel you are using to create the blade and what the hardening temperature of the blade is. Being informed about your alloy's tempering and hardening temperatures is the most essential part of the tempering and hardening process. Do note that if you would like to try out using steel, then you can work with scrap steel before moving on to other forms of steel. We shall cover working with scrap steel further down in the chapter.

During many stages of heat treatment, you can perform any machining or grinding processes that you may want to do. Machining is the process of cutting material to transform it into certain shapes based on your preference. Grinding is another process of adding shape to your material, but this time, you utilize a grinding wheel to make cuts into the metal. You can use any of these processes to create a blade that fits your idea, so now would be a good time to add a little creativity to your work!

When you are producing a blade, one of the most important steps that you will encounter is grinding. Once you have properly ground a blade, you will have an uninterrupted edge that will have little to no flaws. Of course, mastering the grinding technique requires a lot of patience and perhaps more than a few trials and errors. Since grinding is an essential step in bladesmithing, let us look at a few ways to perform the grinding method.

The Hollow Ground Edge

The hollow ground edge has a concave edge. This form of an edge is well suited to blades that will be mainly used for slicing. Examples of such blades include skinners, hunters, filet knives, etc.

The main reason why the hollow ground edge is suitable for such knives is the fact that it produces a very thin edge that can be sharpened quite easily. However, because of this thin edge, the blade can be rather fragile as compared to other forms of grinds. This is why it is not prudent to simply make a hollow ground edge if you are going to be using your blade against heavier substances such as bone, wood, or substances with similar thickness. An important fact to know here is most of the blades produced around the world today are hollow ground. It could be because not many people are looking to cut bone or thicker materials!

Another feature of the hollow ground to note is that it produces a strong and light blade that adopts a sharp edge easily.

Here is how you can achieve a hollow grind.

Step 1: You are going to use a grinding wheel or a belt for a hollow grind.

Step 2: Take the blade and slowly bring the edge to the surface of the wheel of the belt.

Step 3: Now start the wheel of the belt and allow

for the grind to form. For beginners, getting the perfect grind might not be easy. However, with practice you should be able to get the edge that you require.

The Cannel Edge

The cannel edge is also known as the "appleseed" edge. It is a superb option to make an edge for heavy-chopping blades such as cleavers, axes, and anything else that will be used to cut through numerous thick objects such as bones and wood. The cannel ground makes a fairly strong edge and will remain steady for a long time. You might find that the edge is rugged as the edge also features numerous cross-sections. These cross sections can cause the sword to be fairly heavy when you are done working on the blade. Additionally, it is not a difficult edge for you to master. You just have to understand the concept that the entire blade will have a smooth, rounded surface. Once you become aware of this, you will know how to work with the blade and what to do with the grind.

Step 1: Note that this is one of the most challenging edges to master so you should only approach this edge grind if you are confident in your ability to produce it. With just a single mistake, you could ruin your blade entirely. For this grind, you will require a slack-flat belt grinder. This is not a grinder commonly used by hobbyists. Only professionals or industries make use of this grinder.

Step 2: Bring the edge to the surface of the grinder.

Step 3: As the grind is being made, you need to move the blade from the edge to the spine. Essentially, you are creating a convex blade shape that starts sharp near the blade's edge and broadens as you reach the spine.

Step 4: Take sandpaper and use backward strokes (stroking the knife from spine to edge) to perfect the convex shape of your blade. Do note that the longer the blade, the more sandpaper you will require.

The Flat Grind

The flat grind creates a nice balance between the hollow grind and the cannel grind. One of the main advantages it provides is that, since it draws from both the hollow grind and the cannel grind, it has an excellent edge that can bear the brunt of heavy chopping. Additionally, even after multiple chopping sessions, it can still retain its sharpness. However, its main disadvantage is that it is one of the trickier grinding methods to work with so it may take some time to learn.

Step 1: The flat grind is similar to the hollow grind, but is simpler to perform. This is because you are not focusing on the edge alone, but on the entire blade.

Step 2: Your technique involves a process similar to the hollow grind. Bring the edge close to the grinding wheel or belt.

Step 3: When the edge is sharp, continue working on the blade all the way to the spine.

Step: Once you are done, you should notice a linear slope that starts from the edge and goes all the way to the spine.

The Chisel Edge

Some parts of the world use this edge mainly to create various forms of knives. However, in other areas, some of the more common products that feature this edge include carpenter's axes and wood chisels. The chisel edge is a special edge made for a specific reason; it makes the blade sturdy. The chisel edge best fits the manufacturing of tools and utensils.

Step 1: The chisel grind is another easy grind to perform on your blade. The main reason for this is that the blade is flat on one side but has a grind on the other. Start off by bringing one edge of your blade close to the grinding wheel or belt.

Step 2: Do note that you can bring any edge to the blade. Then you follow the same steps as required for the hollow edge. In fact, you can think of the chisel edge as a single-sided hollow edge.

Step 3: Do not worry about how the flat side looks. You do not have to grind the edge in such a way that it appears symmetrical to the other side. This is because there is no edge on the other side to match your grind. Simply grind the edge and sharpen it. The angle of the bevel is entirely up to you and the aesthetic you would like to add to your blade.

Working with 1080

1080 has a somewhat higher manganese composition than other carbon steels in the 10XX category. Because it is a relatively easy steel to work with, it makes 1080 an ideal steel for beginners who want to start their bladesmithing adventure. It gives you enough room to make errors when it comes to heat treatment. It is known to form an almost complete pearlite structure when you subject it to annealing and normalizing processes.

Pearlite is a form of structure that features alternating layers. Because of this arrangement of layers, it is easily considered to be one of the strongest material structures on this planet.

Additionally, 1080 contains nearly 0.80% carbon (which is represented by the 80 in 1080) and is known to produce a good quality knife with a nice edge.

Now comes the rest of the process.

Annealing

In the annealing process, you start off by heating the metal to 1500°F. Then you have to cool the metal but you should avoid cooling it too quickly. You have to ensure that the metal cools at a rate of 50°F per hour or lower. I would not recommend going below 45°F for this purpose.

PRO TIP: In many cases, knife makers use an overnight cooling strategy. For this, you heat the metal to the required temperature of 1500°F at the end of the day. Ensure that the last heat of the day is slowly disappearing by the time you remove the metal from the forge. Once that is done, you then cool the metal in the forge overnight. This becomes handy when you have to perform other work or you might be engaged in the evening.

At this point, you can perform your machining or grinding process, should you wish to.

Normalizing

For the normalizing process, you heat the metal to 1600°F in a forge. You can also raise the heat to a high temperature of around 2150°F. Do not attempt to work on the metal below 1500°F. Once the temperature has been reached, quench the metal for about four minutes. For the right quenching liquid, I have provided a recommendation below.

After four minutes, allow the metal to cool in still air. When you normalize the steel, you are resetting the crystalline structure. Through this reset, you are distributing the carbides in such a manner that they become uniform.

When you are working with steel, having an uneven structure affects its quality. Which is why, if you do not reset the structure, the carbides tend to group together tightly. Due to this, the steel will not have the sharp uniform edge that it could have had.

Quenching

For the process of quenching, you should ideally invest in a fast quench oil. These are special oils that you can use for working with 1080 steel. You need to ensure that the 1080 steel is at

900°F before you introduce it to the quenching process.

PRO TIP: Preheat the oil beforehand. You will be able to cool your steel faster.

Hardening

Your next mission, should you choose to accept it (I recommend you do to get the best results) is hardening the metal. For this, you heat the steel to 1500°F. You are aiming to push it past its non-magnetic limit. In this case, that limit is around 1425°F.

When you are working in the forge, you have to heat the metal until the metal does not attract a magnet to itself. When you have reached such a state, you heat it to a slightly higher temperature. This is just to make sure that you have truly pushed the steel into the non-magnetic area.

If you overheat the steel by keeping it at temperatures of 1550°F or beyond and you quench the metal, the metal could form grains.

Therefore, the best way to complete this process is by heating it to its non-metallic temperature. Then keep it in the forge at that temperature for about a minute. Then remove the steel and quench it. Certain areas of the steel might only require about 1 or 2 seconds of cooling. However, that does not mean that you have to take the steel out of the forge and quickly dip in it liquid. Do not do that! Trust me, that is a safety hazard. Think of it this way.

You take the metal out. You are in such a hurry to beat the 2-second mark that you knock off the oil to the ground. The metal drops and there is a pretty big flare. That flare catches nearby furniture or object that is flammable. Well, you know the rest.

Do not be in a hurry. The steel will hold on to some of the heat and survive for a few seconds when introduced to the air. Take it carefully and place it into the liquid for quenching. Be ready to face a small flare-up along with a high level of smoke.

Tempering

If you have been following the instructions, then your steel should be around 65RC. At this level, it is fairly fragile so do not drop it. It might shatter upon hitting the ground.

Bring the steel to room temperature and begin tempering it once it reaches that temperature. Temper twice. Each tempering process should be done for 2 hours. Allow the steel to return to room temperature between the two processes.

Ideally, your method should follow these steps: temper for 2 hours, then return to room temperature, and then back to tempering.

If you temper at 400°F, then you get a yield of 62RC.

If you temper at 500°F, then the yield is around 60RC.

If you temper at 600°F, then you get a yield of around 57RC.

You should ideally aim to get a yield of 60RC. That would be ideal for working with 1080 steel.

Finally, polish the steel as you see fit.

Working with 1095 steel

Heat Treatment
Once again, working with 1095 steel is pretty simple. It is a steel with high carbon content and you can use it to forge shapes easily. It does have lower traces of manganese than other steel that are part of the 10XX series (such as the 1080 steel.) However, the comparatively higher rate of carbon means that it provides more carbide that can be used for providing resistance to abrasions. However, this also means that because of the extra carbon, you might have to

put in more care during the heat treatment.

Annealing

For the annealing process, you start off by heating the metal to 1475°F. Just like with 1080 steel, you then have to cool the metal at the rate of 50°F per hour. Do not cool the metal any faster. You could also go with the suggestion to cool it overnight. You have to keep the metal inside the forge to ensure the cooling is complete.

At this point, you can perform your machining or grinding process.

Normalizing

To normalize the steel, you have to bring the temperature of the metal to 1575°F. Once the temperature has been reached, you then soak it (or quench it) for about five minutes. After those five minutes, you then allow the metal to cool in air till it reaches room temperature.

Heat and Quenching

For the process of quenching for 1095 steel, you should be using food grade oil. I would recommend using canola oil but any vegetable oil will do. Ensure that the metal is at around 1000°F before you introduce it to the quenching process.

Tempering

The tempering process for 1095 steel is also fairly similar to 1080 steel. However, the difference here is that once you complete quenching, heat the steel to 500°F. Once that is done, the steel will be around 66RC. This will allow it to be fairly brittle but not too much. However, I would still not recommend that you drop it. You might not destroy it entirely but you just might cause cracks to appear on it. Once again, temper it twice for a period of 2 hours each. Make sure that you are allowing it to cool. Bring it down to room temperature before you temper it again.

Cryo Treatment

Now, this step is not entirely necessary. However, it will improve the quality of the steel you are working with. If you like, you can completely skip this step.

If you would like to complete this step, soak the steel in temperatures ranging from -90°F to -290°F. The medium you should choose for cryo treatment should be liquid nitrogen. You need to ensure that you have introduced the metal to cryo treatment for about eight hours. For this, you can even soak the metal in liquid nitrogen overnight.

Polish

Finally, you can add in the polish to give your project that final look.

15N20 Steel

Annealing

The first step that you are going to take is the annealing process. For this you begin by introducing the steel to temperatures from 1400°F to 1450°F. At this temperature, you are making sure that so there is no change in the crystal structure.

To ensure that you are improving the steel's softness, you have to heat the metal slowly. Preferably, heat it to temperatures of about 1500°F, which is about 100°F more than the critical temperature of the metal. At such temperature, the metal enters a phase known as the transformative phase. Now, you have to soak it in liquid, preferably food grade oil, for the appropriate time. At this temperature, the crystalline structure of the steel will start to gain austenite.

Once you soak the steel at the correct temperature it is time to cool the steel. You have to ensure that you cool the metal as slowly as possible. In this case, you should cool it at a rate of 70°F per hour or less. You can even cool the metal in vermiculite. Vermiculite is a type of mineral that you can dip the metal into.

What you should be doing is cooling the metal until it reaches about 100°F below the critical temperature.

Normalizing

When you are normalizing, you follow the very same process as annealing. However, the difference lies in the cooling process. When you are about to cool the steel, then you do not place it in an insulating environment. Rather, you take the piece and leave it in the open air to cool. This is to ensure that the cooling takes place at a much slower rate. When you cool the metal faster, you do not allow the formation of cementite.

And what exactly is cementite? Well, it is the formation of iron carbide crystals in the steel. It is created to improve the hardness of the metal.

Which is why forming cementite is an important step in the process of strengthening your steel.

Heat and Quenching

When you are working with 15N20 steel, you can make use of vegetable or canola oil.

However, when using these oils make sure that you preheat them to at least 150°F. The metal itself should be at least 900°F before you plunge it into the quenching oil.

Tempering

For the tempering process, you have to heat the knife to a temperature of about 1400°F. This brings it to a hardness of around 60 - 65 HRC. When you reach this temperature, you might notice that the knife turns a shade of red. This is important because you are ensuring that you are tempering at the right temperature. Once you have attained the temperature, then make sure you quench the tool immediately. You can use oil for the purpose of quenching. You can even make use of food oils for the quenching process. What kind of food oils you can use are mentioned further down in this section.

Stainless Steel

Heat Treatment

Most of the stainless steel you are going to be using for bladesmithing harden in oil. While some steels can be quenched in preheated oil, others benefit from sub-zero quenching. Just make certain you know the type of stainless steel you are working with before you decide what kind of quenching process you would like to adopt.

For our purposes, we can treat stainless just like we would treat high-carbon steel. That way, you won't face any difficulty when you subject the metal to the hardening and tempering processes. Additionally, high carbon stainless steel is the most useful steel for blademaking.

You have to first heat the stainless steel to a bright, cherry-red color, which typically occurs around 600°F. After that, you quench it in oil. If you notice that you are not getting good results in the oil that you are using, shift the stainless steel to a lighter oil as in some cases (and for some stainless steels) a lighter oil may be necessary.

Annealing

Most of the stainless steel that you get will be in cold-rolled condition. This means that when you subject the stainless steel to annealing, you have to do it at a temperature of 1700°F. Once you reach that temperature, then you should aim to cool it as fast as you can. However, make a note to be very careful when you are cooling the metal. Do not try to shift it from the forge to the medium really fast (as we saw earlier what could happen in such scenarios). When you are cooling it, make sure that you are doing it at

800°F. This means that you have to heat the medium to about 800°F and then dip the metal into the medium.

I would recommend using oil for the purpose of quenching.

When you quench at 800°F, you are allowing the carbon and chromium in the stainless steel to combine. This is great to improve the overall quality of the steel.

You can also use water to quench the metal, but do note that the metal might receive a certain distortion. My recommendation is to try the process on another piece of stainless steel metal (one you are not using for your blade). Check the results that appear on the blade. If you are satisfied with the distortion, then you can choose to use water. If you are not, then you are better off using oil.

Try not to cool the blade in the air. You might notice a slight discoloration on the metal. However, some people prefer to have this discoloration. If you would like to see what it looks like, try it on another piece of stainless steel.

Normalizing

When you are working with stainless steel, you

are typically aiming to improve its strength. When you have increased that strength, you do not want to lose it. Hence, for the process of normalizing, you subject the steel to temperatures of 700°F. If you subject the stainless steel to higher temperatures at this point, then you might unintentionally lower the strength it has gained so far. However, many metallurgists also recommend that normalizing is not required for stainless steel. Here is an important point to remember: if you haven't added stress to the metal, then you do not have to normalize it at all.

Heat and Quenching

When you are using stainless steels, you are using metals that have a have high alloy content. This gives them a high degree of hardness. If you would like to achieve full hardness, then you have to air-cool the stainless steel. However, do note that you are going to receive some sort of discoloration, as we had mentioned above.

If you have to harden large sections, then I recommend that you quench the steel in oil. Steel that is hardened must be tempered immediately after bringing down its temperature to room temperature. This is done in order to prevent the steel from cracking.

Sometimes, the steel is frozen at around -165°F before it is subject to tempering. If you are using martensitic steel, then you have to ensure that you are quenching it at temperatures close to 950°F. For other stainless steel types, you can introduce the metal to the quenching liquid between 850°F and 900°F.

Some stainless steels require more heat treatments than martensitic stainless steel type. However, for the purpose of making blades, the above-mentioned process will suffice.

Tempering

Similar to alloy steels, stainless steels are hardened using tempering. You can temper stainless steel at temperatures of 1800°F. At this temperature, you can attain a hardness between 75 - 80 HRC. Once again, you can quench them in oil once the steel has been removed from the forge. Do note that the different effects that you receive using oil, water, or air. By using the right medium, you can get the desired effect that you are looking for.

O1 Steel

Heat Treatment

The best part about O1 steel? It is so easy to

work with it and make knives. I would recommend beginners to get into bladesmithing with O1 as well. You can easily form shapes with the O1 and once done, you can even sharpen it really easily. You can use the steel to create a nice edge that does not dull easily, even after repeated use.

However (and there is always a however), O1 steel can rust more quickly than other forms of steel. Which is why you must take care that you prevent rust from catching up.

Preventing rust from developing is not a very complicated process. Simply dry and oil the knife after each use.

Now to get started with the forging process for O1 steel. You can ideally forge the metal at temperatures of about 1800°F. You can go lower should you wish to, but ideally you should never forge below the temperature of 1500°F. After you have forged the blade, you then have to follow the below processes.

Annealing

The annealing process used for O1 is done slowly. You need to heat up the metal slowly and uniformly to temperatures of approximately 1140°F. Once the metal reaches the temperature, take it out and soak it in oil. Make sure you soak it thoroughly. After a proper quenching treatment, you then need to cool the metal. Place the metal inside the furnace and let it cool until it reaches a temperature of 1000°F.

Normalizing

For the process of normalizing, you are going to air cool the metal. Begin by first introducing the metal into a furnace and heat it to 1600°F. Once the metal reaches the temperature, remove it from the furnace and then introduce it to still air. That's all there is to it!

Heat and Quenching

For O1 steel, you should utilize oil for quenching. More specifically, ensure that you are working with warm and thin quenching oil. Preheat the oil to 125°F (or as specified in the instructions) before you soak the blade in it. Once you have completed quenching the steel, you have to immediately temper the metal. The steel itself should be at least 900°F before it is dipped into the oil.

Tempering

Now it is time to temper the metal. You need to maintain the temperature at about 400°F, which allows you to achieve a hardness between 56 - 60 HRC. At this temperature, you need to temper it for about an hour to get the results that you require. This also depends on the tool that you are going to create using this steel. Say for example that you are planning on producing a sharp tool, then your tempering process might be a little different. Cutting tools typically require more hardness. To achieve the required hardness, you might have to temper them at a temperature of 350°F for about two hours. Do note that if the sections are under two inches, then you have to temper those sections for two hours. However, if the sections are more than two inches, then you have to temper them for just an hour.

Damascus Steel (1095 and 15N20)

This is a unique steel to work with. As we have already seen, it can produce some beautiful results. This is especially suitable if you are aiming to add a little aesthetic value to the blade. We will be looking at first heat treating the metal and then I will recommend to you the etching stage of the blade as well.

Annealing
For the annealing process, make sure that you heat the blade slowly and evenly to 1500°F. When you reach the temperature, you might notice a dull red color forming on the blade. When you notice this color, then you have to

continue heating it to about 15 - 20 minutes. If you need, you can time the process using a watch or any other device. Being accurate about the timing is key here. You do not want to overheat the metal.

If the blade gets overheated, then it might suffer cracking or warping. And that is something that we are trying to avoid.

Another method of heating is to make sure that the blade goes beyond its magnetic limit. To do this, you have to heat the metal to 1400°F. Once you have reached this temperature, you can test if the metal is ready for quenching. To do this, place a magnet near the metal. If the magnet is attracted by the metal, then you might have to heat the metal some more. If the metal does not attract the magnet, then you can begin the quenching process.

Normalizing

Once you have completed the annealing process, you simply have to take out the blade and leave it out exposed to air.

Heat and Quenching

For the quenching process, you can use either a light or standard quenching oil. However, using an oil quenching method is only strictly necessary for making a large blade where you might need the toughness. Otherwise, you can choose to quench the metal in brine. Quenching the metal in brine is suitable for skinnier blades as you want to ensure they have a nice holding edge. Additionally, you can use brine quenching for small blades. When you use it on smaller blades, it can make the blade harder as compared to oil. However, you need to be careful when you are quenching the blade in brine. If you cool the blade too fast, then you might cause the blade to crack. To ensure that you are quenching properly, simply preheat the brine to around 100°F before you soak the metal in it. You can make a homemade brine solution. All you have to do is dissolve salt in water until you can no longer dissolve any additional salt in the water. When you reach that point, you have a brine solution.

The blade itself can be between 850°F and 950°F before you start the quenching process.

When you are working with thin blades, then you should try to quench either its point or its spine first. This prevents the blade from cracking or warping. When you are working with thick blades, then you can quench it using the cutting edge first. This ensures that you have maximum hardness.

Tempering

We have now reached the process of tempering. For tempering, you need to make sure that you are heating the metal in a heat treatment oven. Ideally, I recommend that you start this process as soon as you have completed quenching. When you make this transfer quick, you are ensuring there are fewer chances of cracking due to residual stresses.

One of the best ways to temper the blade is to take a block or slab of steel. Then heat up the block or slab to a temperature of around 400°F, which should give your metal a hardness between 56 - 60 HRC. Once that is done, place the knife on the steel and allow it to absorb all that heat. You should keep the knife on the steel for as long as you can maintain the heat or at least for one hour.

I would recommend heating it for a period of two hours but this depends on the block and the time you are working with. If you can achieve a one-hour duration of tempering, then that will suffice as well.

Scrap Metal

When you are working with scrap metal, you are making use of recycled steel. Most people think that there is not a lot of heat treatment for this process. That is far from it. There is more to do for scrap metal. Let us dive into it and take a look at the process involved with scrap metal.

Annealing

Since it is scrap metal, it is not easy to find out the temperature at which you would like to subject the metal to annealing. This is why you can use a special tactic. You can just begin at 1500°F. Once the metal reaches the temperature, you can check the results. If you are not satisfied, then try again and raise the temperature to 1550°F. In such a manner, keep increasing the temperature at 50°F increments until you get the desired result. It is alright if you end up heating it a bit more than necessary or at a higher degree than you had planned. As long as you don't melt the steel, you are doing good.

Normalizing

You have to now take the metal into the furnace to normalize it. Preheat the furnace to a temperature of 1300°F. Once that is done, place the scrap metal in the furnace for about one or two hours. I would ideally recommend one hour so that you can check how the steel looks. If you think it requires more normalizing, then, by all means, continue to subject it to the process.

However, if you feel that you have reached a satisfactory level, then you can stop the normalizing process.

Heat and Quenching

In this process, you can take the metal and soak it in oil that is preheated to around 250°F. For scrap steel, try and use fast cooling oils. They have the desired effect and they allow you to retain the hardness of the metal. The metal itself should be around 800°F when introduced to the quenching liquid.

Tempering

For this process, we are going to use a rather unique method. Firstly, you will need to start a fire that can heat the blade evenly. You might also need coal for this process. When you see the fire burning well, break the coal and distribute them evenly into the fire. Once that is done, place the blade on top of the fire in the center. What you are going to do first is heat the blade up as much as possible.

If you would like to know whether the fire is going well or not, just check if the coals are glowing. If they are, then you have yourself a well-lit fire. If they are not, then blast the fire, even more, to make sure that the coals are heating up.

Turn the blade over as many times as you can. This prevents any cracks from forming on the blade. When the blade is preheated, place it in the center of the fire. When the blade is preheated, put it into the center of the fire.

When the blade is in the fire, turn it over every thirty seconds. Move the coals around if required and continue heating. You should be able to see the blade glow red hot. That is an indication that your tempering process is going well. Eventually, when the steel reaches a temperature of around 600°F, you can stop the tempering process.

Chapter 4: The Clean-up Job

Once the knife has been created, it still needs to go through a few more steps. To begin with, we are going to subject it to the final grinding process. What is that? Let me explain it to you.

Final Grind

One of the most important steps that you are going to be taking is the final grind. In order to make this step successful, you should ensure that you are doing this process slowly and carefully.

In all honesty, there are no substantial differences between the final grind and any other grinding process. The steps are actually the same. Perhaps the minor difference lies in the amount of steel that you are planning to remove. Also, you might have to take note of the grit of the belts you are going to use. If you are confused about what I am talking about, then let me explain. When you are grinding, you essentially use belts to smoothen out the knife you created. This helps you remove any excess steel that may be present on the knife.

The main reason for using the final grind is to lower the cutting edge down to a thickness that you can sharpen with. Additionally, you also remove any coarse grit scratches. Now you might need to know how far down you need to take the edge.

My recommendation is that you take the edge down to approximately ten-thousandths of an inch for slicing of cutting tools. For a heavy-duty blade, about fifteen-thousandths. If you are making a sword, then you might need at least twenty-five thousandths of an inch.

If you have a blade and this level of thickness, then you might not have a super sharp edge. However, if you are creating a large blade like a sword (perhaps you would like to cut pineapples?), then you don't really need a sharp edge.

Once again, it all depends on the type of blade that you are aiming to make. If you want a really sharp edge, grind the blade so that you have a thinner edge. If you want a more rugged edge, then grind the blade so that it is thicker. If you would like to know what thickness your knife should be, then consider this: a carving knife has a thickness of about 0.35mm. This ensures that the knife becomes perfect for cutting. For thicker measurements, you can go anywhere from 0.36mm to 0.45mm.

After you begin to notice that the blade is fully heat-treated and tempered, you can begin the final grind. Make sure you grind the blade slowly. Cool the blade frequently. If you notice that the blade begins to change colors, you are destroying the temper. If this continues, then you might end up annealing the blade to the point where it is useless. So prevent the blade from changing colors or getting too hot.

Here is a simple tip to keep in mind. If you find the blade too hot to handle, then that means it actually is too hot. You can reduce the degree of heat that builds up by using newer and sharper belts for the final grinding.

All that you are trying to achieve in the final grading is the same results of the rough grind process. You are also cleaning up the grind lines and ensuring that all of the unnecessary hammer marks are removed. You can do some minor changes as well at this point. Do make sure that you do not overheat the steel while you are performing your change.

If you can manage to grind and polish the flat areas of the blade in advance, then make sure that you do the same for the bevels. You will find it easier to keep the grind lines sharp and clean.

Fitting

After you have completed the final grind, you are now ready to add in the fittings. There are several ways to go about completing this process. You can make them from a sheet, bar, or even a rod of steel or other material. Or alternatively, the fittings can be filed, cast, or forged into shape. Since you are the knife maker, the final decisions are up to you. There is no easy or hard method. It is just a matter of preference.

You might also have to remember that some blades don't require a bolster or other fittings. Examples of the blades that we are talking about are boot knives and hunters.

If you are making a hidden-tang knife, then the best thing you can do is attach the grip material to the tang before you begin to work on the fitting. When you do this, you will be able to fix the grip onto the tang without softening the joint between the guard and the blade.

When you are preparing the fitting for a hidden-tang design, begin with a piece of stock that is a little wider and longer than what you want to end up with.

The guard can be brass, bronze, copper, German silver, stainless steel, or just about anything you have on hand and want to use. Just be certain that the material isn't too hard to drill through and/or file or you will not be able to achieve a good fit on the blade.

Sharpening

Finally, we reach the sharpening stage. When you are sharpening the knife, then it means you have officially reached the last step. In this process, you are trying to put on a good edge to the blade. There are many ways that you can get a sharp edge, the method you choose is up to you.

When you are about to add the first edge on your knife, make sure that you are doing it using a machine. Any further sharpening should be done using a stone. You could also use a leather strap for this purpose.

The main thing that you should focus on is to keep the cutting edge going down the center of the blade's edge. This center portion of the edge is known as the spine of the backbone.

With that, you have completed the cleaning and finishing processes necessary for the knife.

Chapter 5: On the Table

Heat Treatment Temperature Table

Composition	Harden °F	Temper °F	Anneal °F	Norm-alize °F	Quench
Medium Alloy	1700-1800	350-1000	1550-1600	0	Air
High Carbon	1800-1875	400-1000	1600-1650	0	Air
Hot Work Alloy	1825-1875	1000-1200	1550-1650	0	Air

List of Welding Flux

Arc Welding

In the process of arc welding, you are using a power supply and also electrodes in order to perform the welding process. What you are doing is creating an arc between the electrode and the material that you would like to weld. The material that you weld is probably composed of mainly metal as this method is often used in order to fuse metals together. Arc welding is one of the most popular types of welding. But are there any more types? Let us look at some, shall we?

Here is a list of arc welding options that you can use:

- Carbon Arc Welding

- Bare Metal Arc Welding

- Gas Metal Arc Welding

- Plasma Arc Welding

- Atomic Hydrogen Welding

- Shielded Metal Arc Welding/MIG Welding

Chapter 6: Guide to Building a Simple Forge

There's one thing you have to face if you desire to work with metals—heat is needed. With the availability of heat, you can make any metal, no matter how tough, submit to your will. You'll never gain complete mastery over this stubborn material without heat. However, it isn't difficult to finally make the decision to take another step toward teaching yourself this smithing skill.

You'd learn how to build an effective but simple forge in this section. With this forge, you can heat steel hot enough to the temperature needed to effortlessly shape it.

Step 1: Materials

Some rocks, like granite, a bucket of mud, and an iron or steel pipe, are required for building a forge. Due to the fact that heating galvanized metal is bad, do not give consideration to galvanized pipes. However, coating the pipe in some layers of mud and using the forge outside for proper ventilation can be used to counter it.

Step 2: Construction

Place the rocks in a ring. You can create it any size you desire, but it is preferable to keep it small so as to get maximum heat.

After this, you need to use a thick layer of mud to coat the inside of the forge. Coating with mud will prevent the ground from absorbing the heat. The rocks are also insulated by the mud so they don't crack.

Step 3: First Firing

To get the fire sufficiently hot, the metal pipe should be connected to a vacuum cleaner on blow. Some steel can be heated to test how hot it gets and then melted in the can.

Step 4: It Gets Hot!

The forge at this point reaches forging temperatures able to melt steel.

Step 5: Conclusion

The forge can be upgraded a bit and made

bigger.

Chapter 7: Sending Out Metals for Heat Treatment

You have to realize that heat treatment is a fairly long process. You may not be able to achieve the quality that you are looking to achieve, especially if you lack the time to practice. This is because heat treatment requires all your attention, If you believe that you are not capable of doing all the work yourself, then you have other options.

One of those options is to look for outside help.

Let us look at the why's of sending your blades outside.

You do not have all the time to spend on heat treatment. This is a common scenario for many beginners. Sometimes, they might find out while they are working about how much time they have. I would recommend that you try out the process by yourself. If you feel that you can handle it, then you are good to go. If you cannot, then you might have to send it outside.

In some cases, you may not be able to afford all the objects required for the process of heat treatment.

There are always emergencies that you have to attend to. But at the same time, you do not want to let go of your process.

The process might be too complex to master. This is also true for many beginners. Heat treatment is indeed a complex process and even veterans of the trade can make mistakes in their process.

How do you send you blades outside for treatment?

Firstly, you need to identify nearby bladesmiths. You might have to do some research in order to find someone who meets the below criteria:

- They are able to match your price requirements.

- They can produce good-quality products.

Do note that you might not necessarily be able to find a bladesmith who provides inexpensive services and gives you good results.

Here is a list of companies that you can consider for your heat treatment processes (do note that knife maker is headquartered in Canada):

http://www.buckknives.com/resources/pdf/Paul

Bos_Brochure.pdf

http://www.petersheattreat.com/cutlery.html

http://www.texasknife.com/vcom/privacy.php#s ervices

http://www.knifemaker.ca/ (Canada)

Chapter 8: Common Mistakes and How To Avoid Them

Overheating

Whenever you are working with low alloy metals, you can often land yourself into a spot where you can cause overheating and burning. When you work with temperatures higher than 2200°F, then you are likely to cause the destruction of the steel or other forms of low alloy metals. In other cases, you can cause the deterioration of the properties of the metal due to poor administration of the heat treatment.

When you cause such mechanical problems due to heat treatment, then you end up affecting the treated metal's toughness and strength.

How to Avoid Overheating

Firstly, you have to make sure that you are following the steps mentioned in this book. Try and use the temperatures mentioned here. In many cases, knife makers might feel the urge to experiment. This is okay. But you have to make sure that you are taking steps after performing thorough research.

There are also other ways you can ensure that you do not cause overheating. You can check the tools you are using to work with the metals. Ensure that these tools are of high-quality. Be certain that they are protected with anti-decarburizing solutions. If you could not heat the metal evenly, then you should allow it to cool first. Once it has cooled down, you can reheat it again to reach the right temperature.

Using the Wrong Type of Metal

You might think that people could never mistake the type of metal that they are using but you could not be more wrong. Why do such situations occur? Sometimes, when you are stocking materials in your workshop, you can add different kinds of metals. In such cases, it is easy to get confused about the metals you have in your inventory if you don't label them.

You might actually end up using the wrong metal and that could be a disaster. Not only have you wasted material, but you have incurred a loss of time and money on your end. Getting materials is not always cheap.

How Do You Avoid This
Make sure that you have labeled all the

materials you have. Perform an inventory check regularly. Make sure that you are certain of the numbers you have. Always be certain of what you are using. Double check the material before you put it into the furnace. Most importantly, do not be in a hurry to get started. Take your time to make sure you have the right metals and tools.

Wrong Steps

When you temper before the metal is ready for tempering, then you might create a small problem in the process. In order to ensure that you are not making any such mistakes, you have to focus on getting the steps right. In many cases, simply choosing to work with the annealing process without knowing might cause harm. You could cause cracks to form on the surface. You might be none the wiser. That might eventually impact your work.

How to Avoid Them

Make sure that you are making note of the steps before doing them. If you prefer, look at the steps first. See if they make sense to you. If you need to research further, then make sure you do. Look out for expert advice before you start. Once you are absolutely certain of the steps, then you can begin working on the metal.

Quality

It is okay to start off with cheap materials in the beginning. You may not want to risk using an expensive metal before you get a little practice. However, know that by using cheap materials, you might not get the results that you want. You might get an inferior quality product. You might notice cracking on the product. You might even spot some sort of discoloration. These are all expected of a material that is cheap and not sourced with higher-quality standards.

How to Avoid Them

Right. This one is a tricky solution. You see, you might not want to invest in high-quality materials right at the beginning. However, what I recommend is for you to look at your needs. If you are practicing, then it is alright to skimp out on quality. But with that, you might not get the exact results that you are hoping to see. This might throw your observations in another direction. You might not be able to make proper evaluations. So think about it before you decide on what quality of metal you would like to invest in.

Leave A Review?

Throughout the process of writing this book, I have tried to put down as much value and knowledge for the reader as possible. Some things I knew and practice, some others I spent time to research. I hope you found this book to be of benefit to you!

If you liked the book, would you consider leaving a review for it? It would really help my book, and I would be grateful to you for letting other people know that you like it.

Yours Sincerely,

Wes Sander

Conclusion

Heat treatment is a satisfying process. The hard work that you put into it reaps some incredible results. Of course, it all depends on the efforts and the time you put into it.

Do go through this book carefully before you jump in on the heat treatment processes. Make sure you understand the concepts. Most importantly, be careful when you are working with metals.

Always put yourself first. Are you in a safe environment? Are you keeping yourself protected? Are you staying a safe distance from fire and other harmful objects?

Remember, there is no point in trying something when you are not feeling safe.

Another factor that you must consider is that heat treatment is a fairly time-consuming process. I have already mentioned it in Chapter 6 but it is worth mentioning once again. When you are aware of this, you might decide how best to approach the heat treatment process and whether you would like to complete it yourself or seek outside assistance.

Another thing to note is the simple pleasure that you derive out of the whole process of heat treatment. Simply watching your blade come to life is one of the most enjoyable sensations you can experience. I am sincerely hoping that you have such an exquisite feeling yourself as you work with many metals.

So keep yourself protected and enjoy a wonderful heat treatment process.